Searchlight Books™

Spotlight on Climate Change

Climate Change Solutions

Abbe L. Starr

Lerner Publications ◆ Minneapolis

Lerner Publications Company
An imprint of Lerner Publishing Group, Inc.
241 First Avenue North
Minneapolis, MN 55401 USA

For reading levels and more information, look up this title at www.lernerbooks.com.

Main body text set in Adrianna Regular.
Typeface provided by Chank.

Editor: Cole Nelson
Lerner team: Sue Marquis

Library of Congress Cataloging-in-Publication Data

Names: Starr, Abbe L., author.
Title: Climate change solutions / Abbe L. Starr.
Description: Minneapolis : Lerner Publications, [2023] | Series: Searchlight books - spotlight on climate change | Includes bibliographical references and index. | Audience: Ages 8–11 (provided by Lerner Publications) | Audience: Grades 4–6 (provided by Lerner Publications) | Description based on print version record and CIP data provided by publisher; resource not viewed.
Identifiers: LCCN 2021051673 (print) | LCCN 2021051674 (ebook) | ISBN 9781728457932 (library binding) | ISBN 9781728463940 (paperback) | ISBN 9781728461922 (ebook)
Subjects: LCSH: Climatic changes—Juvenile literature.
Classification: LCC QC902.8 (ebook) | LCC QC902.8 .S73 2023 (print) | DDC 363.738/7452 23/eng/20211—dc23

LC record available at https://lccn.loc.gov/2021051673
LC record available at https://lccn.loc.gov/2021051674

Manufactured in the United States of America
1-50821-50160-2/15/2022

Table of Contents

Chapter 1

WHAT IS CLIMATE CHANGE?

We live on the most amazing place in our galaxy: Earth. Take a deep breath. You inhaled oxygen, nitrogen, and carbon dioxide. These gases enable us to breathe and live. We are on the only planet in our solar system with an atmosphere that is known to support life. When the balance of these gases changes over time, so does the weather on Earth. This is climate change.

Earth's atmosphere is filled with many gases and is about 300 miles (480 km) thick. It surrounds Earth and protects it from the sun's powerful heat and radiation. The atmosphere has five layers. They keep Earth both warm and cool.

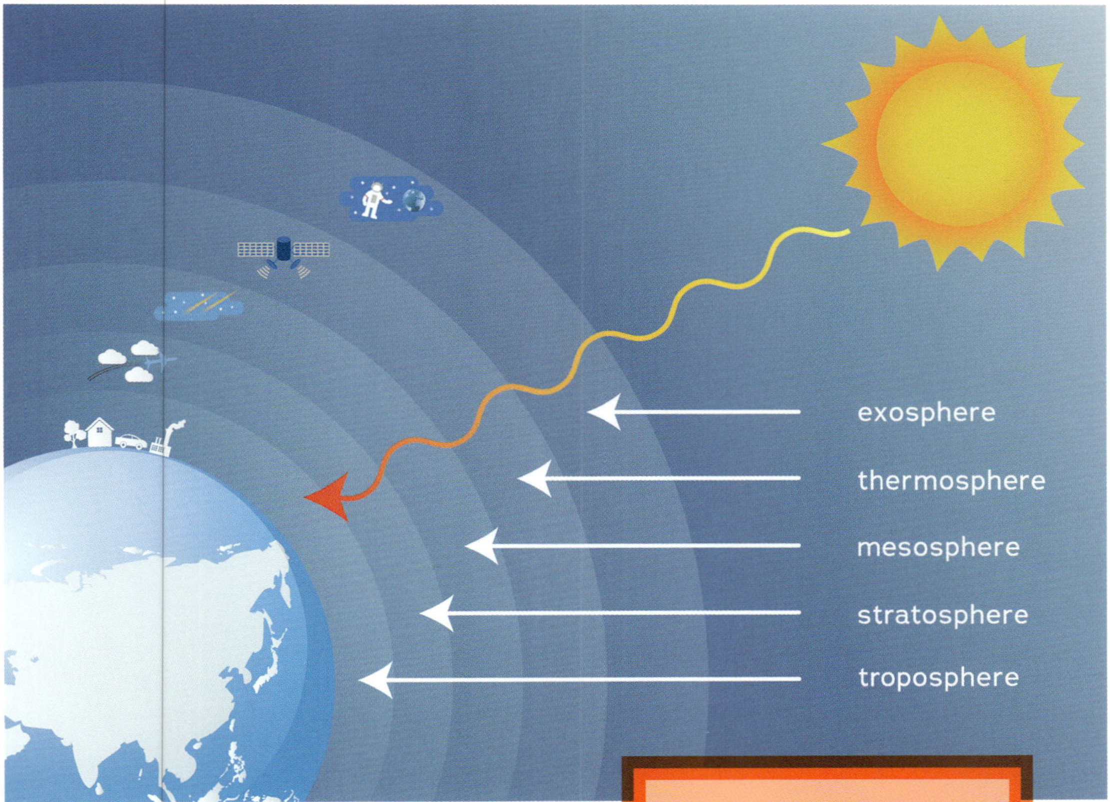

A diagram showing the layers of Earth's atmosphere

Weather and Climate

Inside the atmosphere, a mix of events happens every day. This is the weather. Weather can be different in many regions around the world and can change hourly. Temperature, air pressure, humidity, wind speed, and wind direction can change the weather.

Storm clouds fill the sky above a field in Russia.

The weather is the day-to-day changes in the atmosphere such as rain, wind, and sunshine. The climate is the changes that happen over a long time. Scientists who study the climate take the average temperatures and rains over many years to understand how an area is changing.

The global climate is the average of all areas around the world. Changes in the ocean, on land, from the sun, and in the chemistry of the atmosphere can all change the global climate and make Earth hotter or colder.

RISING TEMPERATURES ARE CAUSING GLACIERS, SUCH AS THOSE IN GREENLAND, TO MELT.

Scientists use satellites in space to collect data about weather and climate events around the world. They discovered that more severe weather, such as hurricanes, is happening more often than ever in human history. Most scientists believe this change is caused by greenhouse gases.

Jason-3 is a satellite that measures Earth's sea levels to learn more about the impact of climate change.

Chapter 2

CAUSES AND CONSEQUENCES OF CLIMATE CHANGE

People depend on the sun every day for energy. In the atmosphere, gases like carbon dioxide capture this energy and keep Earth warm even when the sun isn't shining at night. These gases act like a greenhouse, allowing light to get to Earth and trapping the heat energy inside the atmosphere. They are greenhouse gases.

Earth’s atmosphere has just the right concentration of gases to keep temperature balanced. Without greenhouse gases, Earth would freeze. With too many, it would be too hot for life to exist.

Earth’s neighbors, Mars and Venus, do not have the right balance of greenhouse gases to support life. Mars has a thin atmosphere without many greenhouse gases. Its surface is frozen, and so far, scientists have not found life there. Venus has too many greenhouse gases. Its surface is so hot that it can melt lead.

This is an illustration of the surface of Venus, the hottest planet in our solar system.

A 1771 ENGRAVED ILLUSTRATION OF A BLACKSMITH WORKSHOP

Temperature balance is changing on Earth. In about 1760, the Industrial Revolution began, bringing a change from a farm and craft society to an urban and factory one in Europe and the US. The machines and the fuels to run them have added about 48 percent more carbon dioxide to the atmosphere. As humans burn fossil fuels like coal and gasoline, greenhouse gases are sent into the air. This warms the planet.

Problems with a Warming Planet

Climate change affects Earth in many ways. When Earth becomes warmer, glaciers and ice sheets melt. This adds water to the oceans and causes sea levels to rise. This can cause more flooding, which endangers people, animals, and property near the ocean.

Weather patterns also change as the atmosphere warms up. Extreme weather events like strong hurricanes happen more often as Earth's climate changes. Droughts, floods, and high temperatures affect farmers and food growers.

Temperature changes lead to coral bleaching, which causes coral to lose its bright colors and turn white.

Chapter 3

INVENTIVE SOLUTIONS

Scientists know that reducing greenhouse gas emissions will slow down the effects of global warming. One way to reduce emissions is by not wasting energy. Many homes and businesses are better insulated than they used to be and need less energy to stay warm. When we use less energy, we burn less fossil fuel.

Solar panels convert sunlight into energy. Solar panels have layers of silicon cells. When the element silicon absorbs sunlight, the electrons inside of it begin to move. This movement of electrons within the cells generates electricity. When solar power became a source for power in 1954, it was not very efficient. Modern solar power is much more efficient and continues to improve.

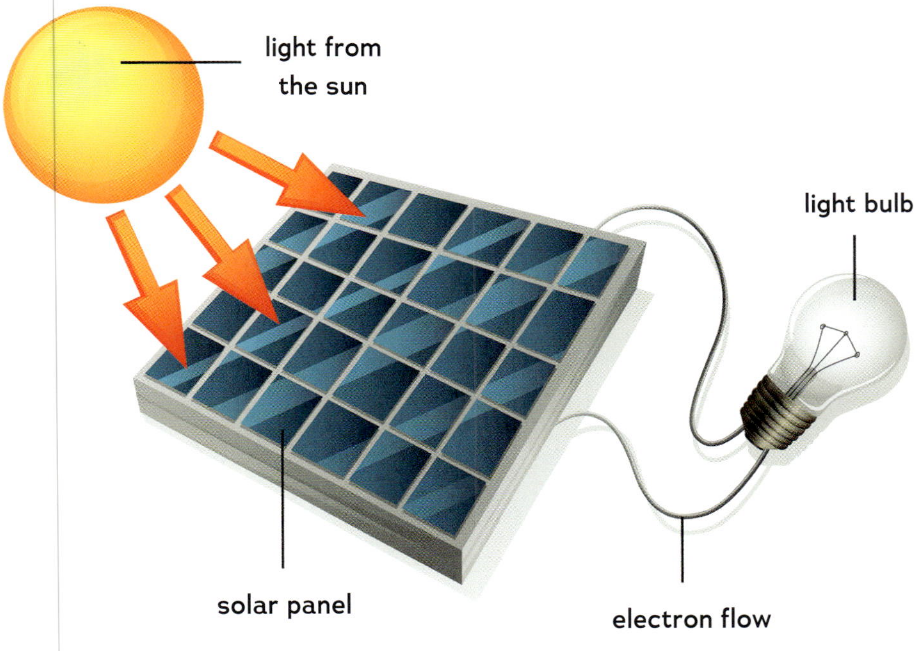

Solar panels capture sunlight to create energy.

Renewable Energy

The biggest way to lower greenhouse gases is to switch from fossil fuels, like coal and natural gas, to renewable energy. Renewable energy is clean and comes from resources that will always be restored like sunlight, water, Earth's heat, and wind.

Solar panels use sunlight as a source of energy.

STEM Spotlight

The first electric car was invented in 1832. Electric cars are environmentally friendly because they emit less fossil fuels than traditional vehicles that use gas. But electric cars take more energy than traditional cars to build. Electric cars also charge their batteries on the electric grid, which uses fossil fuels to get power. Even with these setbacks, the electric car is more energy efficient and better for the environment.

A wind farm in Copenhagen, Denmark

Some forms of renewable energy work by moving a wheel in a generator. This generator creates electricity from that moving wheel. Hydropower creates this energy by using moving water to turn a wheel in a generator. The more water that flows, the more energy the generator can produce.

Wind farms generate power by using the wind to move large blades like a gigantic pinwheel. When the blades turn, so does a rotor that spins a generator. The generator creates electricity that is sent to our homes and businesses.

Geothermal energy uses heat from inside Earth. Wells are drilled deep into the ground to find hot water and steam. This hot water and steam is brought to the surface to turn a wheel in a generator, making electricity.

Nuclear power also uses steam to make electricity. In a nuclear power plant, atoms are split apart into smaller atoms. This releases a lot of energy, which heats water and turns it into steam to run a generator.

A DIAGRAM OF A GEOTHERMAL POWER PLANT

Taking carbon dioxide out of the air is another effective way to fight climate change. As trees and plants grow, they absorb carbon dioxide out of the air. Protecting forests and planting more trees will help keep carbon dioxide out of the atmosphere.

Scientists are searching for ways to improve batteries to save energy. They are also researching ways to capture carbon dioxide and store it in the ground.

Trees absorb carbon dioxide and release oxygen.

Everyone can do things to help the environment and prevent further climate change.

Working for Change

It's not just scientists who are finding ways to prevent further climate change. From changing agricultural practices to inventing zero emission fuels, scientists, farmers, and climate change activists are looking for ways to reduce waste and use resources smartly.

Wangari Maathai

Wangari Maathai (*below, in 2006*) is the first woman in East Africa to earn a PhD and become a professor of veterinary science. In 1977, Maathai started the Green Belt Movement. She worked with women in rural Kenya to help their community by planting trees. Her work protected forests and helped promote the idea of "reduce, reuse, and recycle" in Kenya and around the world. She also worked with the United Nations Environment Programme in its Billion Tree Campaign, which aims to plant a billion trees worldwide.

People can help fight climate change in little ways too. When people need to buy household appliances, they can buy appliances that require less energy to run. People can drive cars that use less gas or run on electricity, making them more energy efficient. Making new appliances and cars uses energy too though, so it can also help if people use them as long as they can. Small efforts like these can make a big difference.

Driving electric cars or carpooling can help the environment.

Chapter 4

LEARNING TO ADAPT

Global climate change is happening and will be an issue for the future. While we can help slow down climate change, people and animals will have to adapt to new weather patterns and environmental dangers.

Environmentalists are protecting and healing wetlands, mangrove forests, coral reefs, and many other natural barriers that protect the shores from waves and erosion.

A seawall in Japan

This helps protect people and animals living near the ocean.

Engineers may also build seawalls to protect cities near oceans.

Many regions will have more flooding as climate change affects weather patterns. City road crews are building streets in new ways that absorb extra water, so roads can stay open during extreme storms.

Other regions will experience more droughts and dry weather. Farmers are adapting to this in many ways. They rotate their crops, keeping soil heathy and full of water, to grow stronger plants.

As weather changes, many animals will need to change their migration patterns or change what they eat. For example, a study has shown that some grizzly bears have switched from eating salmon to eating berries year-round.

When oceans absorb carbon dioxide from the air, they become acidic. This can cause coral reefs to shrink and leave many fish without a home. But lobsters, crabs, and shrimp can build strong shells with higher acidity and can easily find food in a weakened reef.

Many animals, such as bears, may need to change their hunting habits because of climate change.

Robin Wall Kimmerer

Robin Wall Kimmerer is a plant ecologist, professor, and author of *Braiding Sweetgrass: Indigenous Wisdom, Scientific Knowledge, and the Teaching of Plants*. She is also a director of the Center for Native Peoples and the Environment. Kimmerer researches the ecology of mosses and how to protect and grow ecosystems that are in danger of disappearing. She combines her scientific discoveries and her knowledge of Native peoples to support thriving ecosystems. Kimmerer works closely with tribal leaders to restore plants that are native to North America and significant to Indigenous American cultures.

Finding Solutions

Climate change is a big problem, but people around the world are working together to solve it. Scientists, farmers, activists, and many others are teaming up to lower greenhouse gases and create energy and food in ways that don't harm the environment. By lowering your carbon footprint, volunteering to pick up garbage and collect items for recycling, and finding new and creative ways to live sustainably, you can be part of the solution to climate change.

Some people volunteer to plant trees to fight climate change.

You Can Help!

Although climate change seems like a big problem for scientists to solve, we can help solve the problem at home. One way to do this is by looking into ways to use energy more efficiently. Here are some examples:

- Turn off lights and water when not using them.
- Use energy-efficient light bulbs.
- Recycle your garbage when you can.
- Plant more trees.

To make an even bigger difference, you can contact community leaders to make sure your area uses renewable energy. With a little curiosity and care, you too can help invent new solutions to fix climate change.

Glossary

atmosphere: the whole mass of air surrounding Earth or other planets

climate: the weather in regions over a long period

ecologist: a person who studies the interrelationship of organisms and their environments

ecosystem: a biological community of interacting organisms and their physical environment

fossil fuel: a fuel (such as coal, oil, or natural gas) formed in Earth from plant or animal remains

renewable energy: energy from a source that is not depleted when used, such as wind or solar power

solar panel: a device that converts light from the sun into electricity

weather: a mix of events that happen each day in our atmosphere

Learn More

American Museum of Natural History: What Is Climate Change?
https://www.amnh.org/explore/ology/climate-change

Huddleston, Emma. *Adapting to Climate Change*. Minneapolis: Abdo, 2021.

NASA Climate Kids
https://climatekids.nasa.gov

National Geographic Kids: Climate Change
https://kids.nationalgeographic.com/science/article/climate-change

Onuoha, Chinwe. *Climate Change and Life on Earth*. Minneapolis: Lerner Publications, 2019.

Walker, Tracy Sue. *Climate Change Activism*. Minneapolis: Lerner Publications, 2023.

Index

Photo Acknowledgments

Image credits: Crystal-K/Getty Images, p. 5; BestPhotoPlus/Shutterstock.com, pp. 6–7; Monica Bertolazzi/Getty Images, p. 8; NASA/JPL, p. 9; Mark Garlick/Science Photo Library/ Getty Images, p. 11; mashuk/Getty Images, p. 12; Brett Monroe Garner/Getty Images, p. 13; BlueRingMedia/Shutterstock.com, p. 15; Justin Paget/Getty Images, p. 16; 4FR/Getty Images, p. 18; Andreus/Getty Images, p. 19; georgeclerk/Getty Images, p. 20; Jose Luis Pelaez Inc/ Getty Images, p. 21; Stephen Morrison/EPA/Shutterstock.com, p. 22; Dania Maxwell/ Bloomberg/Getty Images, p. 23; Akio Miki JP/Shutterstock.com, p. 25; CarlaMc/Getty Images, p. 26; Mayur Kakade/Getty Images, p. 28.

Cover: zhihao/Getty Images.